Y0-BUC-255

UNDERSEA BASE

BY MAE FREEMAN

Illustrated by JOHN MARDON

UNDERSEA BASE

FRANKLIN WATTS, INC.
NEW YORK | 1974

Also by
the author:
Space Base

Library of Congress Cataloging in Publication Data

Freeman, Mae (Blacker) 1907-
Undersea base.

SUMMARY: Takes the reader to the bottom of the sea to walk underwater, ride in a sea scooter, see a fish ranch, and view aquanauts at work in their undersea base.

1. Manned undersea research stations — Juvenile literature. [1. Manned undersea research stations] I. Mardon, John, illus. II. Title.
GC66.F73 551.4'607 73-9996
ISBN 0-531-02664-7

Printed in the United States of America
6 5 4 3 2 1

UNDERSEA BASE

"*Diving Saucer* ready to leave for Undersea Base 3." These exciting words boom through the waiting room of the shore station. At last you are going to begin your trip to the bottom of the sea.

You have been waiting here for almost two hours. The man sitting next to you is a scientist. He is going to look for some special rocks on the sea bottom. And he is taking you along.

Several other people are in the waiting room. They are going down to work in Undersea Base 3. They pick up their bags and form a line at the door.

The scientist nods to you and stands up. You pick up your bags and follow him. You and he are the last ones in line.

As you go through the door, you can see the *Diving Saucer* that will take you to the undersea base. It looks like a huge, shiny yellow hamburger bun. Most of it is underwater.

There is a narrow bridge from the door to an opening at the top of the *Saucer*. On a ship, an opening like this is called a "hatchway." The people cross the bridge to the hatchway. One after another, they climb down a ladder and disappear inside the *Saucer*.

Then it is your turn. The *Saucer* is rocking in the waves, and you hold on tight as you go down the ladder.

Inside the *Saucer*, the other people are already in their seats. They have kept places near a small window for you and the scientist. The window is underwater, but some sunlight comes through and makes the water look green.

A bell rings. The FASTEN SEAT BELT sign lights up. The hatch cover bangs shut, and the *Saucer* slides away from the bridge. You are on your way to the bottom of the sea.

A person who works under the sea is called an "aquanaut," just as people who go out into space are called "astronauts." So now you can call yourselves junior aquanauts.

The *Saucer* moves slowly away from the shore. Then it begins to move downward, too.

Through the window, you can see that there is not as much light as before. Hundreds of silvery shapes dart past the window as the *Saucer* passes through a school of small fish.

After a while, you hear a steady, hissing noise. It is the sound of air coming into the *Saucer* from a tank. There is a very special reason for this.

The air around you presses hard on everything, even on your bodies. But you do not notice this pressure, because there is air in your lungs, and even in your blood, that pushes outward just as hard.

Water presses on everything in it, the same way that air does. The deeper the water, the harder it presses.

The *Saucer* is going downward, so more and more air is pumped in from a tank under the floor. This keeps the air pressure inside the *Saucer* the same as the water pressure all around it, so the *Saucer* will not be crushed. The change happens very slowly.

By the time you get to the undersea base, the air pressure in your bodies will match the water pressure down there. And you will be quite comfortable.

It will take much longer to come back up because the pressure on your bodies will have to be lowered again. This takes many hours. That is why an undersea base is so useful. People can stay there and work for many days without having to come back up.

Now the *Diving Saucer* is about halfway to the undersea base. There is much less light than before, and the water looks blue. The powerful headlights are turned on, but the light does not reach very far.

You keep your faces close to the window, trying to see through the blue water. But suddenly you jerk backward! Outside, a big head is pushing up against the window. The head has two bulgy eyes and a mouth like an angry bulldog.

It is only a harmless fish called a "grouper." The grouper does not seem to be afraid. It nuzzles the window for a moment, then turns and moves slowly away.

Other fish nearby seem to have no fear either, as the *Saucer* moves along. Imagine how it must look from the outside—a huge, yellow monster with two big, bright eyes.

All at once, the *Saucer* tips sideward as it makes a turn. Your seat belts keep you from falling.

Now, through the window, you can see the undersea base. It looks like a huge, metal balloon resting on tall, steel legs. There is a steel dome under one end of the base. The dome looks like a big, upside-down cereal bowl.

When the *Saucer* gets closer to the dome, you can look upward and see a big opening in the bottom of the base. You can even see bright lights inside.

With such an opening, you would think the whole sea might rush in. But it cannot, because the air inside the base is at the same pressure as the water all around. So the air cannot leak out and the water cannot flow in. The people in the base are actually inside a huge air bubble.

It takes only a minute for the *Saucer* to wiggle into place. When it fits tightly in the bowl, the hatchway is inside the base while the rest of the *Saucer* is still in the water.

Then the hatchway opens and a ladder comes down from the base. One by one, the people in the *Saucer* climb up. When it is your turn, the scientist gives you a boost—and here you are, in the control room of Undersea Base 3.

On the wall of the control room there are all kinds of dials and TV screens. Colored lights flash here and there.

An aquanaut is sitting at a desk that faces the wall. He is wearing earphones as he watches the dials and the lights. From time to time, he pushes a button, or turns a switch, or makes notes in a book.

The base has two floors. Right above the control room is the engine room. It is a noisy place. Motors are humming, pumps are clicking, heaters are hissing, water is gurgling.

No one is in the engine room right now. But the aquanaut in the control room is watching the signals that show how all the machines are working.

Next to the control room there is a small kitchen. You are glad to go in there because you are feeling a little hungry. Breakfast seems a long time ago, up on the shore.

At home, food can be cooked in open pots and pans. But in the base, this would put too much steam into the air. So all food is cooked and frozen on shore. Then it is brought down here and kept in a freezer. The frozen packages of food are warmed by putting them into a radar oven for a few minutes.

After lunch, you start on your visit to the rest of the base. In the next room, you tiptoe very quietly. The lights are dim, and two aquanauts are asleep in their bunks. In a few hours, they will go back on duty, and other aquanauts will have their turn to sleep.

Now you go into the most interesting place of all. It is the science workroom. The aquanauts call it the lab.

Several people are at work in different parts of the lab. One of them shows you his animals. He has some tiny white mice in small cages on a shelf. In a bigger cage, a few young rabbits are nibbling lettuce. Two small, friendly monkeys are swinging from a pole.

All these animals have been living in the base for a long time. The aquanauts take good care of them. They want to see if animals can stay healthy when they grow up in high-pressure air.

On the other side of the lab, there is a big tank with hundreds of odd-looking fish in it. Some sea plants are growing in the sand at the bottom of the tank.

On a steel workbench, there is a live baby dolphin. It is held in a kind of plastic crib with straps to keep it from sliding off.

The dolphin has been put to sleep with chemicals, and an aquanaut is measuring its breathing and its heart beat. When this work is finished, the baby dolphin will be taken to a big aquarium on shore.

When this baby grows up, it may be one of those you see doing tricks in a water circus.

The scientist who brought you here now shows you his own part of the lab. On the table, there are all sorts of tools, bottles, and chemicals. Behind the table, on a shelf, there are rows of plastic boxes. Some of them are filled with sands of different colors. Others have rocks in them, or stones, or shells.

The scientist picks up a smooth-looking stone and shows you how to put it under the microscope. Now you see that the stone is not really smooth. It has many rough places and some tiny cracks. A few specks sparkle here and there.

There are useful minerals in some of these rocks. They are used to make steel, cement, and many other things. Some day, when there are no more of these minerals left on land, they will have to come from the bottom of the sea.

Now you have a pretty good idea of what people do in an undersea base. All over the world, scientists like these are working—under the sea, on the earth, and out in space, too.

The afternoon goes quickly in the lab because there are many interesting things to see and to do. Then it is time for dinner. It is fun to sit at the table with other aquanauts and hear about their adventures.

Afterward, you are quite ready for bed. As you tumble into your bunks, you try to stay awake for a while to think about the exciting day. But you fall asleep and dream about tomorrow. For tomorrow you are going out for your first undersea walk.

A walk on the sea bottom can be dangerous. It is quite dark out there. The water is very cold. And you must take along air to breathe.

For safety, aquanauts never go out alone. When two aquanauts go out together, they are called a "buddy pair."

The get-ready room is called the "wet room." That is where you and your buddy get dressed.

First, there is a rubber suit that zips up tight. It is heated by electric wires all through it. A strong plastic suit fits over the rubber suit to protect it.

Next, you put on gloves, a helmet, a face mask, and heavy boots. Then a tank of air for breathing is strapped to your back.

The tank is the main part of your SCUBA outfit. Scuba is a word made up of the first letters in

Self-**C**ontained **U**nderwater **B**reathing **A**pparatus.

Last of all, electric batteries are hooked up to your outfits. They will run the heating wires in your rubber suits and the searchlights on your helmets.

You are ready to go—but you can hardly move. All the things you are carrying are very heavy. But when you get into the water, you will not mind them. They will seem much lighter because the pressure of the water will help hold them up.

Your buddy starts down a ladder into the water, and you follow him. As soon as you are underwater, your loads seems lighter.

At the bottom, you hold onto the ladder tightly to keep your balance as you look around. You are in a big, wire cage. It was put there to keep sharks away from the hatchway.

The door of the cage is open, and you try to walk through it. When they are going short distances, aquanauts find walking easier than swimming. But walking down here is much harder than walking at home. Down here, you have to push against the water.

You work hard to catch up to the rubber suit moving ahead. With each step, you seem almost to float a little, as astronauts do on the moon. When astronauts are in training, they walk underwater. It helps them get ready for moon walking.

Not far from the base, two aquanauts are setting up a big tank that has a TV camera in it. A thick plastic tube with electric wires inside runs from the tank to the base.

A worker in the base will be able to make the camera turn all the way around, or tilt up and down. In this way, he can watch all the sea animals that happen to come near.

There are several TV cameras like this one in other places on the sea bottom.

The men working on the tank have a good way of getting the tools they need. One of them has a small buzzer hooked to his suit. He presses the buzzer and in a few moments, a big shape zooms out of the darkness. It is a trained dolphin, bringing a bag of tools from a supply ship floating up above.

The aquanaut takes the tool bag from the dolphin's mouth and gives the animal a friendly pat. It turns around, flips its tail and zooms off again. It has just heard the buzzer on the supply ship, calling it back. Dolphins can hear this sound through the water, but people cannot.

No one knows why dolphins can dive deep and come up again very quickly without harm from changes in water pressure. People are not able to do this. They must come up very slowly while the pressure is lowered.

Dolphins are easy to train. This one is used as a messenger because it was trained to swim toward the signal from a buzzer. It can also be sent out to find a lost worker and lead him back to the base.

Now you and your buddy will go off to hunt for rocks. This means a ride in a sea scooter. The scooter is a little car shaped like a bullet. It is very useful, because walking under water is not easy.

The scooter does not really scoot. It moves very slowly, but it is better than walking or swimming. When a giant sea turtle swims by, it moves faster than the scooter.

Some long tubes are hooked to the side of the scooter. They are used for cutting chunks of earth from the sea bottom. A tube is pushed down into the sea bottom and then pulled up again. It comes out filled with layers of mud and sand.

Back in the lab, the layers are taken out of the tubes. The chunks of earth show how the sea bottom has changed over thousands of years.

Sometimes, interesting things are found in the layers—tiny sea animals and even bits of gold.

Your next visit is to an underwater ranch. It is not at all like a ranch up on land where horses or sheep are kept. The ranch down here is for fishes.

The fish ranch has a fence that is quite special, because it is made of air bubbles. Long pipes are set in the shape of a big square on the sea bottom. There are holes all along the pipes. Pumps push air through the pipes, and it bubbles upward from the holes. This makes a wall of bubbles.

The cowboys on this ranch are dolphins. Two aquanauts in a scooter give buzzer signals. The dolphins follow the signals and drive small fish through an opening in the bubble wall. The bubble wall keeps the fish in one place so they can be gathered by a fishing ship that floats up above.

In this strange, new world under water, things sometimes have tricky names. For instance, a fish ranch is not very much like a sheep ranch. A dolphin cowboy is quite different from a cowboy out west. There is even a forest down here. But it is not at all like a forest up on land.

An underwater forest is a place where only huge vines grow. They have long leaves that move gently in the water. Many kinds of fish dart in and out among the vines, like birds flying around in a forest of trees.

These vines are called "giant kelp." They are a kind of seaweed that grows upward toward the light. The kelp grows so tall that men in boats up above can gather it.

This kind of seaweed is very useful. There is a chemical in the leaves that is used for making paint, medicine, and even candy.

There is one more interesting visit to make before you leave Undersea Base 3 for home—a scooter ride to the wreck of a pirate ship that sank more than two hundred years ago. Some aquanauts discovered it when the base was built.

Your scooter moves slowly around the wreck. It is an old, wooden ship, black and broken, and covered with mud. One of the cannons is still there, all rusted and cracked.

Maybe there is treasure still hidden somewhere. Your buddy parks the scooter, and you crawl around on the wreck. When you scrape away some mud near the cannon, you find old gold coins.

Everything found by aquanauts on the sea bottom belongs to the base. But you are special junior aquanauts, and so you can keep the coins.

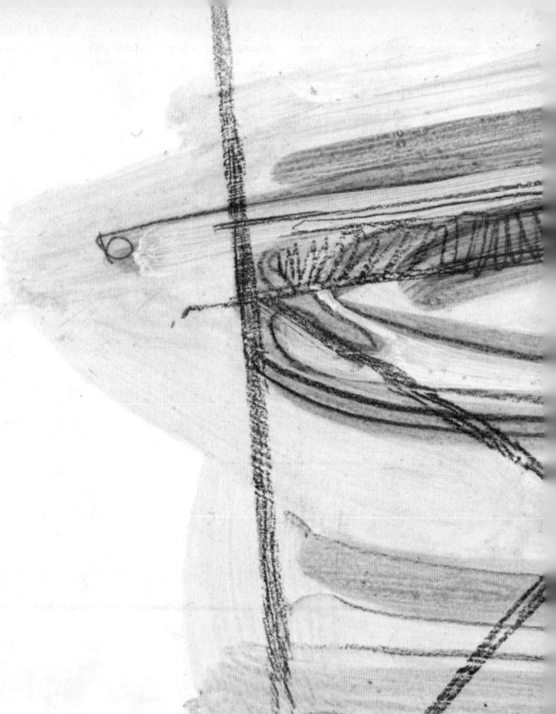

The time comes when you must leave Undersea Base 3 to go back home. You go down through the hatchway again and into the *Diving Saucer*. Some other aquanauts are leaving the base, too.

You will not be able to get out of the *Saucer* as soon as you get back to the shore station. Your bodies are now used to the high pressure under water. It would be dangerous if the pressure dropped too quickly.

So you will have to make another stop before you can go out into the air.

When the *Saucer* reaches the shore, it fits into another dome. This time, the hatchway leads into a pressure tank. The tank is just a small room without windows.

You and the aquanauts must stay in this tank while the air pressure is dropped very slowly. It will take several hours.

You do not notice the change, and you are quite comfortable. When the pressure in the tank matches the pressure outdoors, you can go out safely.

And that is when your trip to the bottom of the sea is really over.

ABOUT THE AUTHOR

Mae Freeman is a native of Chicago, Illinois, and a graduate of the University of Chicago. She has written a number of books for children, mainly in the field of science, and is the author of Space Base, *the companion volume to* Undersea Base.

Mrs. Freeman now lives in Bound Brook, New Jersey.